D0590477

The Secret Life of an Oakwood

The Secret Life of an Oakwood

A Photographic Essay

STEPHEN DALTON

with Jill Bailey

CENTURY

London   Melbourne   Auckland   Johannesburg

Dedicated to 'The Dragon'

This edition first published in 1986 by
Century Hutchinson Ltd,
Brookmount House, 62–65 Chandos Place,
Covent Garden, London WC2N 4NW

**Century Hutchinson Publishing Group
(Australia) Pty Ltd,**
16–22 Church Street, Hawthorn, Melbourne,
Victoria 3122

Century Hutchinson Group (NZ) Ltd,
32–34 View Road, PO Box 40–086, Glenfield, Auckland 10

Century Hutchinson Group (SA) Pty Ltd,
PO Box 337, Bergvlei 2012, South Africa

Photographs and captions © 1986 Stephen Dalton
Text © 1986 Jill Bailey

All rights reserved. No part of this publication may be
reproduced or transmitted in any form or by any means,
electronic or mechanical, including photocopy, recording
or any information storage and retrieval system, without
permission in writing from John Calmann and King Ltd

Stephen Dalton's photographs are available from NHPA,
Little Tye, High Street, Ardingly, Sussex

Dalton, Stephen
   The secret life of an oakwood.
   1. Oak—Great Britain   2. Habitat
   (Ecology)
   I. Title   II. Bailey, Jill
574.5′2642   QK495.F14

ISBN 0 7126 1278 5

The author would like to thank Keith Boyer for his
enthusiastic help in locating some of the plants and
animals which appear within these pages

This book was designed and produced by
John Calmann and King Ltd, London
Designer Richard Foenander

Typeset by Composing Operations Ltd, Tunbridge Wells
Printed in Hong Kong by Mandarin Offset International Ltd

Contents

7
Introduction

17
Spring

63
Summer

107
Autumn

141
Winter

158
Notes on the Photography

159
Index

INTRODUCTION

he oak, with its large, spreading crown, deeply furrowed trunk and twisting roots, is surely the most majestic English tree. Leaning against a mature oak, several centuries old, one can feel that sense of power and timelessness that emanates from ancient monuments, be they great cathedrals or mighty trees.

From the ancient Greeks to the North American Indians, from the Norsemen to the Druids, the oak has featured in mystical religion and folklore. Even today, the focus of many English villages is an ancient oak on a village green. In its shade people sit and talk, and on summer evenings, youths park their motorbikes against its trunk. Notices pinned to its bark tell of fêtes and fairs, marriages and funerals, items lost and found, and other announcements of local import.

Like the village oaks, forest oaks are the focus of life around them. The shade cast by their crowns determines the pattern of shrub and herb growth below. In spring their young buds and pollen-rich catkins feed bullfinches and tits and thousands of caterpillars. Beetles tunnel through the bark, and birds nest among the branches. The acorn harvest supports jays, squirrels, mice, and in ancient times wild boar, through the long winter. Hedgehogs, mice and adders hibernate under the spreading roots, and the rotting leaves provide food for fungi, earthworms, snails, slugs, and millions of tiny soil organisms. Minerals drawn up by the roots from deep in the soil are incorporated into the leaves. When they fall and decay, the minerals are released to shallower layers of soil, there to be taken up by herbs and shrubs.

To walk in an English oakwood is to tread the path of centuries of history, for the woodland has been shaped by man for thousands of years. The story of the oak started ten thousand years ago, when the great ice-sheets melted and the landscape emerged into the sunshine. Trees began to invade the tundra, spreading north through the continent, of which Britain was still a part. As the climate warmed, more and more species invaded, until by 7,000 BC woodland covered all but the poorest soils and the steepest slopes.

This was the ancient wildwood, as yet little penetrated by man. It was not uniform, but a mosaic of different kinds of wood, with different species thriving on different soils, a mosaic that slowly changed as the various species competed for light and space. The oak arrived around 7,000 BC, and by the time the

English Channel was formed, around 5,000 BC, cutting off Britain from the continent, oak was probably the second commonest tree in the land after lime.

The ancient oakwoods were in many ways similar to those in Britain and on the continent today. In summer the leafy crowns of the mature oaks and a few other large trees such as beech and lime formed an almost unbroken layer across the top of the forest: the canopy. Beneath them the shade was deepened by the understorey – the smaller trees of the forest – hazel and birch, holly and yew, and the saplings of taller trees. These grew densest where the oak canopy was thin, perhaps where one of the giant trees had blown down during a winter storm. Lower still was the shrub layer, tangled thickets of hawthorn and bramble, and in sunny glades dog rose, blackthorn and guelder rose, often interwoven with climbers: honeysuckle, traveller's joy and bryony. On the forest floor grew the herbs, a changing patchwork of colour: anemones, violets, primroses, celandines and others. Here and there grassy glades brought splashes of sunlight into the green underworld, where grazing deer prevented the seedlings of the forest trees from growing. Butterflies flitted in and out of the sunshine and fawns lay invisible in the long grass.

Yet the wildwood was not quite the same as a twentieth-century oakwood. There were no great carpets of spring flowers: anemones, bluebells, daffodils, and wild garlic. The oaks were closer together and much taller, perhaps 90 feet (30 metres) or more. The whole scene was more untidy, with shrubs straggling up towards the understorey and a mixture of small trees of many different heights. There were more tall herbs, and fewer flushes of low-growing flowers such as the primrose.

What makes the modern oakwoods so open is their history of coppicing. In many species of trees, when the main trunk is cut off close to the ground, new shoots sprout from the stump or stool, producing several main shoots growing close together. Because they shade each other, they grow tall and straight in their race to the light, producing uniform sets of poles which have many uses in farms and gardens, fences and gates, and other constructions. Many of the hazels in the understorey of the oakwood, although quite slender, are offshoots of stools that may be hundreds of years old. When the poles were cut, it was customary to clear the undergrowth at the same time, so that there were no tall herbs and shrubs to shade the new growth. This also gave the lower-growing herbs of the woodland floor a chance to grow, and the first and second springs after coppicing saw carpets of anemones, bluebells and wild garlic, celandines and daffodils growing among the old stools. Gradually, taller plants shaded them out, but they remained as underground bulbs and corms, swollen roots and seeds, to burst out again at the next coppicing. In Neolithic times, about 3,000 BC onwards, the clearing of forest to create farmland began in earnest. There was an increasing

need for wood for farm implements, bridges, gates, fences and houses, and fewer woods to provide it, so woodland management was important. With the arrival of the Iron Age, around 500 BC, charcoal was in demand for iron production, and the wealth created by the iron-based industries led to an increase in population. The late Iron Age and the Roman period (AD 40–400) saw the greatest rate of forest clearance in English history: over half the wildwood was felled in just seven hundred years.

But settlements were not confined to the cleared land: until only a few centuries ago, the woodland supported a human population of its own. The oakwood was not the quiet, deserted place it is today. Mingled with the birdsong was the sound of human voices, dogs barking, the metallic thud of axes, and the rumble of carts trundling firewood along the muddy tracks. One might even hear the lowing of cattle in the woodland glades. Wood pastures, kept open and grassy by grazing, were quite common. Houses were made of wattle-and-daub – panels of woven twigs supported on timber frames and daubed with clay or mud – and were easily erected wherever a man could find work in the forest. The old woodland industries called for many different specialists – woodmen, sawyers, carpenters, hurdle-makers, basket-weavers, charcoal-burners and woodmongers – and their products were often sold to villages and towns many miles away.

The rights to hunt different beasts for food were often allocated according to social status, for this was still a feudal society. Nobles hunted deer, mostly red deer. Lesser men could hunt roe deer, martens, foxes, hares and wild cats. Falconry was popular, and many small birds were eaten. Woodcock were netted as they flew along the rides at dusk.

In the autumn, pigs were led into the forest to feed on the acorns. Hazelnuts were so abundant that whole families travelled from far and wide to camp in the woods for weeks on end, collecting the nuts to sell in neighbouring towns. Sloes were an important crop, sold to chemists or used to adulterate port wine.

By Anglo-Saxon times, from the sixth century AD onwards, rights of access to the forest were carefully regulated, and rights to feed pigs on the acorns, collect different types of wood, fell timber, hunt, and pasture cattle were strictly controlled. The King and his court, who moved around the country helping to keep order and preside at criminal trials, expected to have a ready supply of game in most forests. Many jobs were created to supervise the King's hunt and to manage the wood for the deer.

Throughout history oakwoods have provided jobs for many, and raw materials for new industries: charcoal for the iron industry, ash for glass-making, bark for tanning leather, cogs for machinery, and frames for printing presses. Oaks have supplied the massive timbers for great cathedrals, and for the great armadas which have defended nations and carried men to colonize new lands.

Today the woods are almost deserted, the woodman's skills forgotten, the neatly coppiced understorey often reduced to a tangle of thickets. Yet there is a growing demand for firewood once again, as the rising cost of energy increases the popularity of wood-burning stoves. Conservationists are trying to revive the old practices of woodland management in order to restore the oakwoods to their former state. Patches of woodland are jealously preserved as valuable parts of the nation's heritage. Perhaps it is hardly surprising that today, too, people return to the places where their ancestors lived, places dominated by ancient oaks that despite the passage of centuries retain their strength, dignity and beauty. They come for peace, to enjoy a wild place away from the roar of traffic and the demands of city life. They come to walk and to wonder, to listen to the birds or just to pass a few hours in quiet contemplation. The oakwoods are places for spiritual renewal, where the outside world can be forgotten and the inner world can slip back into perspective.

One such place is a small oakwood near Ardingly in Sussex, in south-eastern England: a piece of ancient woodland, once part of a forest which in Roman times extended for almost 90 miles (150 kilometres) from Kent to Hampshire. There has probably been an oakwood on this site since the time of the ancient wildwood: the presence of small-leaved lime trees and certain other plants indicates a long history. The trees here would have been used by the Romans for their iron-smelting industry, by the Saxons for their houses, by the Normans to support high church roofs, and by countless generations since for building houses and ships. Today few of the oaks are more than a couple of hundred years old, but they are still impressive, dominating the woodland landscape. There are patches of hazel coppice, and along woodland streams, alder trees grow among banks of Sphagnum moss and saxifrage. Now it is managed as a nature reserve.

This book provides an insight into the life of this oakwood: not the human activity of the Middle Ages or the passage of twentieth-century weekenders, but the secret life that belongs to the oakwood alone, the plants and animals, seen and unseen, that for thousands of years have lived, died and been reborn in its shadow, under its protection.

SPRING

resh green turf sparkles with dew in the glades and rides, and threads of beaded gossamer sway in the early morning breeze. Beneath the hazel catkins, the stream bubbles amidst banks of yellow saxifrage, and thick cushions of moss glow among the wet rocks. From high in the canopy, blackbirds and thrushes proclaim the arrival of spring; below them, wrens sing loudly from perches in the tangle of last year's bramble.

From the furthest reaches of Europe and Africa, birds are arriving, drawn to the woodlands by some instinct, each new arrival adding to the chorus. On the first warm breezes of spring, chiffchaffs blow in, their songs echoing from the treetops. Soon the notes of blackcaps and garden warblers pour from the greening hollows, and willow warblers fill the trees with melody.

Beneath the furrowed forms of ancient oaks, a carpet of wood anemones spreads across the woodland floor. Within the circle of arching roots, bells of wood sorrel nod among their trefoil leaves, and raise their flowers towards the strengthening light. In shady corners the tiny moschatel wafts its scent over nearby swathes of dog's mercury.

The oaks awaken. A yellow haze creeps through the canopy as catkins slip from shiny buds and waver in the wind. The leaves will stay in bud longer, and before they shade the woodland floor, a mist of bluebells hides the ground, scenting the warming air. Primroses lie scattered around the trunks of trees.

As young leaves reach out towards the light, an army of insects emerges from a million tiny eggs in tree and soil. Caterpillars feed on the new spring foliage, eagerly sought by sharp eyes and probing beaks. Green tortrix moth caterpillars lie safe within a rolled-up oak leaf, carefully fastened with silken threads. They are so numerous that even the largest trees may be almost stripped of leaves, to stand as stark reminders of winter. If danger threatens, they drop on fine threads to the ground. Other larvae feed by night, resting by day rigid as twigs and brown as bark. Deep within oaks and beeches, beetle larvae munch crazy

---

An **oak** seedling (*Quercus robur*) sprouts on the woodland floor, threatened by deer and rabbits hungry after the long winter. It will grow at most 12 inches (30 centimetres) in its first year, a tiny fraction of the height of its mighty parent. Over two or three hundred years the oak can grow to 82 feet (25 metres), and 24½ feet (7½ metres) in girth.

tunnels through the wood, and giant goat moth larvae lie unseen in the decaying timber.

From buried chrysalids and pupae, other flowers of the forest escape. The first butterflies flap along the rides or bask, wings spread, in the warm sun. Along stitchwort-studded paths, orange-tips pause to sip nectar, and sleepy brimstones cling leaf-like to slender bluebell chimes. Columns of midges dance in shafts of light, and large bumblebees buzz across the woodland glades, searching for old holes or burrows in which to rear their broods.

There is no shortage of takers for the growing food supply; for now is the season of birth and growth, and the northern summer is short. Courtship is in full swing among the treetops, and soon there will be new mouths to feed. Birds which through the winter months hunted in solitude or crowded in noisy flocks, now look for mates and stake out territories, anxious to control a large enough share of the feast to feed their growing families.

The hardships of winter forgotten, squirrels scramble and scold through the trees on their courtship chase. With a clap of wings, the wood pigeon bursts out of the canopy in an exuberant courtship flight. Throughout the wood, smaller birds posture, sing, chase and display. By night, the pursuit continues. The badger summons his mate with strange purring commands, while the hedgehog scampers on short legs after a reluctant female. Angry squeaks in the long grass signal the start of matrimonial disputes among the smaller mammals.

Deep in thorn-bound thickets, among the leafy twigs or low in the long grass, birds are building nests, flying to and fro with grasses, twigs, lichens and leaves. On the woodland floor the woodcock sits on her eggs, her stripes and speckles mingling with the sun-dappled leaves in perfect camouflage. In burrows under the sprawling roots and tunnels deep in the brown leaf litter, mice and shrews are busy with the first young of the year. The tiny naked bodies, with tightly closed eyes, jostle on the teats of their warm mother.

Beneath the roots the adder stirs, and slithers sleepily into the sun, slowly stretching its diamond coils. Down by the stream, the grass snake slides through clumps of yellow kingcups, searching for the frogs so noisily announcing their amorous intentions. A brown rat scampers along an overhanging branch. The dormouse sleeps on, waiting for darkness before coming out to feed.

At night, the earthy smell of moist dead leaves wafts up towards the new leaves overhead. The wood is alive with the sounds and movements of animals.

The **willow tit** (*Parus montanus*) can be distinguished from its very similar relative, the marsh tit (*Parus palustris*), by its larger and duller black crown. The springtime nests of willow tits are most likely to be found in the damper areas of the wood where there are rotten tree stumps in which the birds can excavate their holes.

With their feathered enemies asleep, caterpillars emerge to feed, and snails glide over the damp leaf litter. Voles and field mice forage among the growing herbs of the forest floor, and the dormouse swings in the slender hazel twigs, eating the delicate new growth.

But the hunters are also awake. They too have growing families to feed. The badger cubs, born naked and blind during the last weeks of winter, are now strong enough to venture out with their mother, exploring the smells of the spring nights, sampling the worms now so near the surface, and probing new nooks and crannies. The weasel sneaks silently through the undergrowth, its beady eyes alert for every movement. High in the oak, fluffy-coated owlets blink their dark, liquid eyes as they wait for supper.

The warmer days of spring are often cooled by showers. The feathery treetops glow under the steel-grey clouds, and the wood grows strangely silent as the patter of raindrops advances through the trees. Glistening droplets slide downwards as young leaves bow before the onslaught, falling through the scented air to bounce again on the trembling herbs below. Water snakes down the furrowed bark of the oak, carrying minerals to the lichen fronds, and soaking the dark mosses that fringe the massive roots.

After the rain comes the fragrance of newly washed earth and flowers, the damp musk of regeneration typical of spring. The water washes minerals deeper, to the very anchors of the forest trees. Seeds swell and burst, shoots push upwards through the soft soil, and pairs of round green leaves unfold in sun-touched corners. Around the trunk felled by the winter storms, seedlings start their race to the light.

The young green of spring deepens, grass grows lush in the clearings and shoots of bracken curl upwards along the rides. Red campions deck the sunny paths, and may blossoms drift across the glades. Safe in the long grass, the newborn roe deer fawn waits motionless, its large ears directed towards every small rustle in the grass. The fox cubs are taking their first faltering steps outside the den, ears and eyes alert for their mother's return. Still in baby fur, they yelp and squabble over the spoils. Already the nightingale's song rings out from the nearby thicket, and down the woodland rides ripples the bubbling cry of the female cuckoo as she lures the male toward her. Summer has arrived.

The Daddy-long-legs or **crane-fly** (family Tipulidae) is one of the most recognizable insects, as it dances off with legs dangling. There are several hundred species, this being one of the smallest; it is on the wing during spring. The larvae of some of the larger species are commonly called leatherjackets; they live in the soil and eat the roots of plants.

The graceful, nodding white flowers of the **wood anemone** (*Anemone nemorosa*), with their conspicuous yellow anthers, carpet many woodlands during spring (below). On dull, cold days, the flowers close and droop to protect the pollen. The ancient Greeks called this the 'wind-flower', believing that only when the wind blew did it open its petals, as if it wanted to be caressed by the breeze.

Above this starry floor, clouds of yellow pollen are released from dangling **hazel** catkins (*Corylus avellana*) or 'lambs' tails' as they swing in the wind (right). These catkins, which are male, start forming in the autumn and open very early in the spring. The female catkins are small and squat, with bright crimson styles, and usually ripen after the male catkins to prevent self-pollination. Although only a small tree, the hazel predominates in the shade of many mixed woods, providing an abundance of food in the form of leaves and nuts for numerous mammals, birds and insects.

One of the commonest and most familiar of birds, the **wren** (*Troglodytes troglodytes*) may be found in any area of dense vegetation, woodland being a favourite haunt (left). Its brown plumage, up-tilted tail and mouse-like way of creeping about low down in the undergrowth make it unmistakable. For such a tiny bird, both its song and its alarm calls are surprisingly loud and penetrating. It nests any time between April and July, laying up to twelve white, faintly speckled eggs.

New leaves contrast with the old as a **bramble** (*Rubus fruticosus*) springs into life (right, above). It is in its shelter that the wren has built its beautifully domed nest of moss, grass, leaves and feathers, with an entrance hole at the side (left).

The lovely **brimstone butterfly** (*Gonepteryx rhamni*) is a harbinger of spring, the sulphur-yellow males waking up from hibernation during the first warm April days. The paler greenish-white females appear a week or two later, and at a distance can easily be confused with the cabbage white (*Pieris brassicae*), but the brimstone has a more direct flight. Although typically a woodland insect, it also flies around hedgerows and gardens. The caterpillar feeds on buckthorn (*Rhamnus catharticus*), and has a habit of resting along the midrib of a leaf, quite camouflaged.

Perhaps the most glorious time of year is April–May, when the **bluebells** (*Hyacinthoides non-scriptus*) begin to come into flower. It is difficult to reconcile this scene of exquisite beauty, vibrant with life, with the scenes of man's destruction elsewhere. The buds are bursting into leaf in the warm sun; insects are on the wing and the sound of birdsong fills the air.

Throughout spring, **seedlings** push up through the dense leaf litter to find rare patches of sunlight (above). Only in the gaps not overshadowed by mature trees do they stand a chance of growing. All over the woodland floor, thousands of seeds lie dormant, waiting for one of the giant trees to fall and allow the life-giving light to reach them.

The moist, shady floor of deciduous woodland provides an ideal habitat for a wide variety of **ferns**, some woods supporting them in almost tropical abundance (family Polypodiaceae, right, above).

Known sometimes as 'the lady of the woods', the **silver birch** (*Betula pendula*) is one of the first trees to break into leaf in spring. Here the roots run over the sandstone surface before penetrating the soil (right, below). It is common in acid oak woodlands and, like the oak, provides food for a multitude of insects as well as for animals further up the food chain.

Also known as the cuckoo flower, the **lady's smock** (*Cardamine pratensis*) shows its graceful flowers at about the same time as the cuckoo in April (left). The plant is equally at home in moist meadows and in damp woods.

During March and April, the large buttercup-like marigold or **kingcup** (*Caltha palustris*) makes a brilliant splash of gold in wet woodland areas and stream margins (right). The glossy, orange-yellow flowers are often over an inch (2½ centimetres) in diameter. It is common and widespread in marshy places.

The **grass snake** (*Natrix natrix*) is most often encountered around the borders of woods and hedgerows, particularly if there is water nearby (left). Its favourite food is frogs, but it will readily take newts, fish and mice. A grass snake can be distinguished from an adder by its more gracefully tapering body and the yellow patch at the back of the neck, forming a collar.

The low, creeping, mossy mats of the yellow-leaved **golden saxifrage** (*Chrysosplenium oppositifolium*) are a vivid sight when it first appears in March (right). It too is common in damp, shady places, especially around springs and woodland streams.

Fortunately, it is not necessary to sit up all night to glimpse this magnificent animal since, unlike the badger (*Meles meles*), the **fox** (*Vulpes vulpes*) often prowls during the daytime, particularly when it has cubs to feed – usually about April (above, left and above). Unlike the badger too, the fox does not have a permanent home but tends to move from one earth to another, especially when disturbed. Country foxes feed mostly on birds, insects and mammals, especially rabbits, which are usually stalked; but a fox may also catch rabbits by a kind of hypnosis. It attracts their attention by playing some distance away, gradually drawing closer while the mesmerized rabbits continue to gaze at the performance; then it pounces.

Were it not for the shooting man, one would rarely catch sight of the handsome **pheasant** (*Phasianus colchicus*), with its multitude of iridescent colours, wandering around the woods and hedgerows (left, below). With its short, stubby wings, the pheasant is designed for rapid take-off; its long tail helps it to weave in and out between the trees.

A dead **chestnut** leaf (*Aesculus hippocastanum*) has held on obstinately throughout the winter, but now that spring has arrived it has little time left (right).

A sun-seeking **honeysuckle** (*Lonicera periclymenum*) spreads its new leaves to absorb the soft spring light (below). As other, taller trees put out their leaves, it will twine around the neighbouring twigs and stems, climbing ever higher to keep one step ahead in the race towards the sun. Its sweet-scented flower will not appear until May at the earliest.

It is exciting coming face to face with this shy beast while wandering through woodland (right). Preferring dense cover to open countryside, the **roe deer** (*Capreolus capreolus*) spends most of the day in hiding, venturing out only in the evening in small parties to forage, and then keeping to definite territories. The roe is one of the mammals in which delayed implantation of the embryo occurs. Shortly after the egg is fertilized, growth is suspended for about five months, until December when development is resumed.

The **jackdaw** (*Corvus monedula*) breeds not only in woods, but also in town centres and around sea cliffs (above). This bird has found an old walnut tree at the edge of the wood in which to rear its young; it lays between three and six eggs in April. Jackdaws tend to be gregarious, and like all the crow family are highly intelligent. They have a complex system of communication, using a variety of different calls and attitudes, and can often be seen, heads cocked, watching intently some small happening in the wood.

In a deciduous wood, a wealth of wild flowers gain a footing wherever they can. In April, when the tree canopy is still leafless, plants such as wood sorrel (*Oxalis acetosella*), primrose (*Primula vulgaris*) and wood anemone (*Anemone nemorosa*) come into flower. As the shadow of the developing canopy spreads over the woodland floor during May, more shade-resistant flowers appear: dog's mercury (*Mercurialis perennis*) and, of course, **bluebells** (*Hyacinthoides non-scriptus*).

The **green-veined white** (*Pieris napi*) is most often seen from April to June, flying around marshy meadows and open countryside, but it can also be found in open woodland (left). The yellowish scales concentrated on the veins of the underwings distinguish it from other members of its family. The caterpillar feeds on plants related to the cabbage, such as hedge mustard (*Sisymbrium officinale*), garlic mustard (*Alliaria petiolata*) and charlock (*Sinapis arvensis*).

The early morning dew brings evidence of other, more secretive inhabitants of the woodland. Oakwoods host a myriad species of spiders, some lying in ambush for insects in the crevices of the bark, others cocooned among the dead leaves, still others, such as the **money spider** (family Linyphiidae) spinning delicate snares on the slenderest of grass blades (below).

The **beech** seedling (*Fagus sylvatica*) typically braves the world by bursting through the leaf litter under its parent tree (below). It is unlikely to grow much larger before being eaten by some woodland creature.

The **beech** is one of the most magnificent trees (right). Its massive, smooth, grey-green trunk, the vast but beautifully contoured limbs, the purity of its spring foliage and its gorgeous autumn colours endow the tree with a majesty that few others can match.

The sleepy, soft cooing of the **wood pigeon** (*Columba palumbus*) is one of the most soothing sounds of spring and summer days (above). It is much larger than other doves or pigeons, and is distinguished further by the white mark on its neck and the white bar on the wing, which is conspicuous in flight. The long breeding season extends from April to October, and occasionally into winter. This bird is carrying a twig for its nest.

The **red campion** (*Silene dioica*) is one of the most familiar flowers of the countryside, decorating woods, hedge-banks, sea cliffs and mountainsides (right). It will flower well into summer.

Oakwoods support a richer variety
of mosses than other types of
woodland, and the **feather moss**
(*Thuidium tamariscinum*) is especially
attractive, with its bright green
branch stems growing in thick,
feathery tufts. It can be found in
damp, shady places around rotting
trees.

This delicately tinged, pale green perennial is widespread and common in shady woodland areas. The mauve-veined white flowers of the **wood sorrel** (*Oxalis acetosella*) are borne on leafless stalks, while the leaves, on shorter stalks, fold up at night. The flowers are light-sensitive, and country folk claim to tell the time of day by the degree to which they have opened.

During April and May, the deeply cleft, delicate white flowers of this **stitchwort** (*Stellaria holostea*) contrast with the surrounding spring green, forming starry splashes in woodland and hedgerow (below). The flower, when mixed with powdered acorns and wine, was once thought to cure a pain in the side.

The rapid, fluttering flight of the **orange-tip** (*Anthocharis cardamines*) as it chases through woodland clearing, hedgerow and meadow is a sight familiar to country dwellers in May and June (right). Only the male has orange patches on the forewings. The eggs are laid on flowers of lady's smock (*Cardamine pratensis*) and garlic mustard (*Alliaria petiolata*); the caterpillars feed on the seed pods.

Although the **blue-tit** (*Parus caeruleus*) is a frequent visitor to gardens during both summer and winter, it is really a woodland bird (left). Gardens may be safe places for the young, but there is usually a shortage of food. It is easier for the parents to rear large families in deciduous woods, as nestlings can consume up to one thousand caterpillars each day. Woodland clutches are therefore generally larger than those of the garden birds. Blue-tits nest any time from mid-April to late June, depending on the availability of food.

Like the green oak-roller (*Tortrix viridana*), the **green longhorn** (*Adela viridella*) belongs to the Microlepidoptera, a large group of small moths (right). The moth is a bronzy-green and has extremely long antennae, particularly in the males. On sunny spring days the males can be seen dancing in courtship swarms around trees: these were photographed at the top of a field maple. The larvae feed by burrowing between the upper and lower surfaces of the leaves of oak or birch.

The **wood mouse** or long-tailed field mouse (*Apodemus sylvaticus*) is an extremely supple and agile rodent, capable of running, jumping and climbing with consummate skill (below). It needs to be, for it has many enemies, including owls, weasels, foxes and even hedgehogs. The success of the wood mouse is due to its semi-nocturnal habit, its great adaptability, versatile behaviour and variable diet.

Also known as ramsons, the **wood garlic** (*Allium ursinum*) bears some resemblance to lily-of-the-valley (*Convallaria majalis*). This widespread plant grows in profusion in and around damp woods, permeating the air with its heavy, garlicky scent (right). Like the bluebell, it is a member of the lily family.

of downy feathers, they wait anxiously to be fed. From time to time they launch themselves into the air, flap frantically for a few yards, then, their efforts abandoned, crash-land.

The oak is under attack. Generations of caterpillars come and go as the season progresses. Many have already feasted on the new spring growth, and pupae now lie buried in the ground or cling, camouflaged, to the bark of the trunk. Green tortrix pupae lie wrapped in silk in rolled-up leaves. These larvae have consumed so much foliage that the oak is putting out new leaves to compensate: the green canopy is studded with reddish-pink shoots, sprouting from buds which were dominant through the spring. Tough and tannin-rich, these leaves present a new challenge to the army of caterpillars.

The beeches have completed their leafy green mosaic, and little flourishes in the deep shade below. Here the fallow deer fawn lies, its white-speckled body blending with the dappled sunlight on the russet leaves. On a lichen-covered branch of oak the nightjar rests, vigilant but motionless. The laughing cry of the green woodpecker reverberates as it dips from tree to tree.

The roe deer are in rut. Around a circle of open ground the buck marks out an area, fraying young trees and scraping the ground with his antlers, marking the boundary with scent from special glands. Encouraged by the excited squeaking of a receptive female, he chases her round and round, their hoofs wearing a track on the ground.

At dusk, the birdsong starts again as darkness ends the search for food; the approaching night provokes a last proclamation of territorial rights. Hawk moths hover over the honeysuckle, probing the deep trumpets. High up in the trees, little green oak tortrix moths search for mates. Deer wander through the glades, browsing on the lush grass, their white rumps visible through the grey of evening.

With the soft churring of the nightjar, night closes in. Among the dark trees, bats tumble and turn in pursuit of moths, and pick off insects sleeping in the foliage. Hedgehog families root among the leaves on the woodland floor for grubs and beetles. The badgers are also active, searching for worms, shrews, mice and hedgehogs. A stoat snakes across a clearing, and the scream of a rabbit breaks the silence.

It is high summer; from a distance the oakwood shimmers in the heat. Now that courtship and the rearing of young are almost over, the birds are quieter, and many are moulting to winter plumage. Already some have left for their winter

---

The **jay** (*Garrulus glandarius*) is rarely found far from trees, where it spends much of its time hunting for small animals, eggs and nestlings. Especially during early spring, when many nests are not well concealed amidst the foliage, jays may play havoc with the young songbird population.

homes: the nightingale has gone, and the turtle dove too. In the clearings, grasshoppers chirp in the grass, and crab spiders wait in ambush on the bramble flowers for butterflies to alight.

More frequent storms signal the changing of the seasons. Thunderclouds tower above the forest and lightning strikes into the canopy, seeking out the tallest crowns. Leaf-laden branches tear and snap in the wind, and dying flowers are beaten to the ground by the lashing rain. This is part of nature's cycle: the dry ground soaks up the water, feeding it to the fruits now swelling on bushes and trees.

Already the oaks have acquired a bronze sheen; the beeches have lost the unblemished green of spring, and the canopy casts a deeper shade as the days shorten and the sun sinks in the sky. Autumn colours are beginning to appear: hawthorn berries, blackberries and pale hazel cobs. Strings of bryony berries twist between the bushes, reflecting the mellow light.

Oak apples are conspicuous among the branches, some green and fleshy, others hard and hollow. Formed around the larvae of tiny wasps which laid their eggs on the oak in spring, they swell as the grubs inside them grow. When they are mature, male and female wasps emerge to start the cycle again. Other wasps cause other galls: marble galls, round and woody, and knopper galls, brown furrowed caps for autumn acorns. The leaves too have their galls, reddish spangles that ripen and fall, carrying their occupants with them.

The summer songbirds are restless. Warblers, chiffchaffs, blackcaps, flycatchers will soon follow the retreating sun south. The winter residents, their summer moult complete, start to feed in earnest for the lean months ahead. Blackbirds listen for worms for their still-growing brood, and wood pigeons collect twigs to tidy up their nests in readiness for a final batch of eggs.

The nights grow colder, and in the weakening early morning sunshine, mist curls out of damp hollows and dew glistens on grasses and herbs. All through the wood, tiny spiders float on gossamer threads, carried by faint breezes to new homes. Among them drift the earliest seeds of autumn, thistledown and parachutes of willowherb. From the treetops, the first yellowing leaf breaks free and spirals to the ground.

The germander or **bird's-eye speedwell** (*Veronica chamaedrys*) is a hairy, delicate-looking plant with flowers of brilliant azure and a whitish eye. It is a lover of grassy places and hedge-banks, and is often found brightening up woodland rides and glades. Its English name, 'germander', means 'ground oak', so called because its leaves are shaped rather like oak leaves.

Beautifully marked in delicate shades of olive-green and pink, the **elephant hawk** (*Deilephila elpenor*) is a magnificent insect (left, above and below). One may be lucky enough to glimpse it around nightfall in June hovering over flowers, probing them for nectar with its long proboscis. It lives wherever the larva's food plant, the willowherb, grows in abundance, in open woodland areas and on city wastelands. Its name derives from the brown tapering larva, which resembles an elephant's trunk.

During late summer the caterpillar of the **privet hawk** (*Sphinx ligustri*) can be found feeding on privet and ash: both may grow in deciduous forests (right). The horn on the final segment is a characteristic of hawk moth larvae. This is one of the larger European hawk moths, and as well as being found in the countryside, it is often seen in the middle of towns, where it visits nectar-bearing flowers.

Like the majority of the warblers, the **blackcap** (*Sylvia atricapella*) is much more often heard than seen, its mellow warbling song being one of the most beautiful sounds of the countryside (above). Until recently the bird was considered a summer visitor to northern Europe, but it has now been established that some stay during winter, supplementing their normally insectivorous diet with berries, and food from bird tables. Sadly though a large number of these delightful songbirds are slaughtered in some Mediterranean countries while on migration.

Seen on heaths, moors and hillsides and in woodland clearings, the **adder** (*Vipera berus*) does not like the hot summer sun, but prefers to lie in the shade (right). It is quite distinct from the grass snake, having a stumpy appearance and rarely exceeding 8 inches (20 centimetres) in length. This striking reptile has a ground colour which varies from brick-red to black, normally with a distinct dark zigzag marking down the length of the back, although plain, unmarked specimens are not uncommon.

The **great spotted woodpecker** (*Dendrocopos major*) can be distinguished from its cousin, the smaller and scarcer lesser spotted woodpecker (*Dendrocopos minor*), by the red patch under the tail (left). The greater spotted prefers woods where there are at least some mature trees, although it will penetrate agricultural land which has not been levelled by the removal of trees and hedgerows. To attract a mate and proclaim its territory, the male drums by tapping its beak in rapid succession against a suitably resonant tree or branch – one of the more haunting of spring sounds. Now that summer has arrived, the parents are working hard to feed their young in a hollow tree.

In midsummer the beauty of this simple flower can be seen in sunny places throughout the countryside and beside woodland paths and glades. The **dog rose** (*Rosa canina*) has been immortalized in many a carved church pew or heraldic coat-of-arms, a reminder of gentle fragrance and soft summer sun (above).

A lover of oakwoods, the **jay** (*Garrulus glandarius*) is a shy, restless and noisy bird. It is difficult to observe it for long: often all one sees is a rear glimpse as it disappears into the foliage, when the white patch above the black tail is very prominent. A closer view will show that this handsome pinkish-brown bird has a chequered patch of azure blue on the wings, and a crest on the head, which it raises at will. Jays are long-lived: some have been known to reach the age of fourteen. They also tend to remain in a small area, rarely flying more than a mile or two from their birthplace.

With a subtle iridescent sheen on its thin membranous wings, the **giant lacewing** (*Osmylus fulvicephalus*) is a jewel among insects (left and below). During daylight hours it rests on the underside of leaves of the dense vegetation bordering woodland streams. The insect flies mostly at night, but if disturbed during the day, its characteristic slow, floppy flight is unmistakable. The eggs are laid on stream-side plants and the larvae live amongst the wet moss around the banks, feeding on the larvae of midges and other small insects. This insect is by no means common, being restricted to certain areas where it may be locally plentiful.

The **ragged robin** (*Lychnis flos-cuculi*) is similar to the red campion (*Silene dioica*) in both form and colour, but is more slender, less hairy and has untidy red petals (right). Although widespread and common, it is confined to wet woods, marshes and fens. The Latin name means both 'lamp' and 'cuckoo flower', the latter because it flowers at the time the cuckoo arrives, between April and July.

To many people, orchids are special flowers, and a number can be found in woodlands. A typically woodland species is the **common helleborine** (*Epipactis helleborine*), which grows in dark and dingy corners of the forest, flowering well into September (left). The unscented flowers are greenish-yellow to purplish and grow up to about 8 inches (20 centimetres) high.

The **spotted orchis** (*Dactylorhiza fuchsii*) flowers in midsummer and is the commonest orchid in Britain (right). It flourishes in many habitats, from wooded places to downland and marshes, always avoiding acid soils. It grows either singly or in large groups, and varies widely in both size and colour.

In midsummer the showy flower-heads of the **guelder rose** (*Viburnum opulus*) are noticeable from some distance (left). The large, conspicuous outer flowers are quite sterile; the greenish mass of small flowers in the centre are the ones that will produce the brilliant scarlet berries in autumn, often in such profusion that the slender twigs droop under the weight.

The **common agrion** (*Agrion virgo*) is an insect more reminiscent of the tropical rainforests than of a Sussex oakwood (below). The favourite haunts of these exotic damselflies are the fast-flowing, stony-bottomed streams of woods such as this. During courtship in May and June, dozens can be seen whirring in the dappled sunlight like miniature helicopters – a magical sight. Only the male has the flashing, iridescent, deep-blue and green wings; those of the female are smoky brown. The green or brownish nymphs creep among the aquatic weeds in search of any living creature they can find. In winter they hibernate at the stream bottom, usually taking two years before they emerge.

In high summer, wherever woodland rides are in flower with thistles or bugle, the pearl-bordered (*Boloria euphrosyne*) or **small pearl-bordered fritillary** (*Boloria selene*) can be seen (above). They are very similar, the small pearl-bordered preferring wetter conditions.

The **sparrowhawk** (*Accipiter nisus*) is the supreme master at woodland hunting (right); it flies swiftly, silently and low, suddenly dropping on its unsuspecting victim. Its short, rounded wings make for great manoeuvrability as it swoops through the tangle of the forest. Usually it is inconspicuous, resting in the cover of trees.

The **honeysuckle** (*Lonicera periclymenum*) is a familiar woodland climber, often found inseparably entwined with the guelder rose (below). Its attractive and fragrant flower has made it a popular plant for the garden.

It is always exciting to watch this graceful and noble butterfly, the **white admiral** (*Limenitis camilla*), flying effortlessly around a woodland clearing or glade in search of honeysuckle or bramble blossoms (right). When freshly emerged, these insects are a velvety black or dark brown, but after a week or two of flying round the bramble thickets, they tend to become ragged – not that this appears to affect their aerial skills. In July the female lays her eggs singly on honeysuckle, and when the larvae are half-grown in autumn they hibernate for the winter, to resume feeding in the spring.

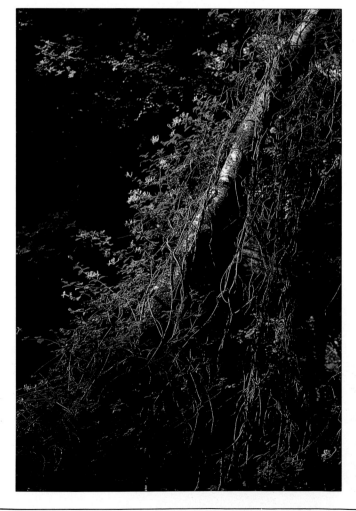

Each year, the little **turtle dove** (*Streptopelia turtur*)
flies all the way from West Africa to raise its offspring
(above). Its soft, soothing, purring notes, one of the most
evocative of countryside sounds, can be heard floating
down woodland rides on balmy spring and summer days.
The bird will breed both in woods and in more open
countryside.

Hazy in the heat of a sultry afternoon in midsummer, the
oaks (*Quercus robur*) are glowing in the late sun (right).
Already among the leaves, the acorns are swelling, and the
buds of next year's shoots are forming in their protective
sheaths of scales.

Sometimes known as the 'drooping sedge', the **pendulous sedge** (*Carex pendula*) is one of the most distinctive sedges, with smooth, willowy stems growing up to 5 feet (1½ metres) high (left, below).

From early spring to midsummer the glowing, yellow-green flowers of the **wood spurge** (*Euphorbia amygdaloides*) are one of the splendours of damper woods, bursting out from among the spiralling red leaves at the sides of woodland paths (right, below). As in all spurges, the male and female blooms are separate, a ring of male flowers surrounding each female one.

The **sulphur polypore** or chicken-of-the-woods (*Lactiporus sulphureus*) is a large, fan-shaped bracket fungus which usually grows in tiered groups on deciduous trees, oak being the most popular habitat (right). It can be found any time between late spring and autumn, and when eaten young is considered a great delicacy in France and North America.

An enchanting little trunk-creeping bird, the **nuthatch** (*Sitta europaea*) is omnivorous in its feeding habits (left). As well as taking insects, the bird feeds on acorns and other nuts, which it breaks open by wedging them in a crevice in the bark before splitting them with its strong beak. Nuthatches are almost unique in cementing up their nest entrance hole with mud until it is exactly the right size, as seen here.

Longhorn beetles are long and narrow, with long legs and particularly long antennae. The larvae feed inside wood, while the adults – if they eat at all – feed on nectar. A common and active beetle in summer, the **spotted longhorn** (*Strangalia maculata*) visits flowers in and around woods (below).

Very common in damp woods from May onwards, the slender, hairless form of the **yellow pimpernel** (*Lysimachia nemorum*) creeps low over areas exposed to bright light (below). The flowers, supported by thin stalks, open in fine weather.

As well as using echo location for catching insects on the wing, the **long-eared bat** (*Plecotus auritus*) can locate prey passively by listening to insects rustling in the foliage (right). Here its long, sensitive ears are directed towards a tasty morsel on an **oak** branch (*Quercus robur*). Sadly, bats are becoming increasingly scarce, partly because man is gradually eliminating their favourite roosts. Dead and hollow trees are cut down, and the animals are discouraged from roosting in lofts – a tragedy, as this much maligned creature is among the most fascinating, friendly and beautiful of animals.

AUTUMN

Autumn drifts in, almost imperceptibly: a thickening of evening mist in the hollows, a glistening of dew along the rides and glades, a glint of gossamer threads in the understorey. The late summer silence is broken only by the loosening of leaves and the scurrying of creatures in the undergrowth.

Fruits and seeds are swelling and ripening, filling with sugars made from the summer sun. Slowly the harvest grows: glossy hips of dog rose, drooping berried heads of guelder rose, blackberries, sloes. Above them hang the nuts: pale hazel cobs in ragged cups; chestnuts in spiky green cases; and acorns, thousands of them, hanging from the branches, hidden in the leaf litter, waiting to be collected and stored for a banquet in the scarcity of winter.

The oaks are lords of the woodland. Without the autumn acorns, many small animals would perish during the winter months. All over the wood, caches of acorns lie, shielded from prying eyes under leaves, or in grass-lined nests under the very roots of their great parents.

Day and night the harvest continues. Nuthatches creep up and down the trees, wedging their nuts in the deep-creviced bark while they hammer them open. Jays carry off beakfuls of acorns, burying them all over the wood and often outside it too. Somehow they find them later, even beneath a blanket of snow. Squirrels argue as they scamper up and down the branches, chiselling at the shells with their long incisors until they split them clean in half.

In the woodland underworld, bank voles blunder through the tangle of withering plants, their heads held so high to keep the nuts aloft that they cannot see where they are going. Out of sight, they grasp the nuts in their forefeet and gnaw off the pointed end to reach the tasty kernel. At night, short-tailed field voles move in from the surrounding fields, lured by the prospect of an easy meal.

The oak can afford its generosity: of the millions of acorns it produces over several centuries, only one needs establish and grow to maturity to replace it.

This striking fungus attaches itself to the roots of a birch tree: the fungus aids the progress of soil nutrients into the roots, while the tree supplies sugar to the fungus. In common with other *Amanita* species, the **fly agaric** (*Amanita muscaria*) is highly poisonous, causing intoxication, hallucination and sometimes death.

But an oakling cannot grow under the shade of its parent, or compete for nutrients with the massive spreading roots of the old tree, so the oak must disperse its acorns widely to find a place to germinate. For this it needs the woodland animals, who carry off the nuts and bury them, often forgetting them until it is too late and spring arrives.

Floating in the damp autumn air, thousands of tiny seeds, borne on silken parachutes, drift through the glades and swirl up into the canopy: willowherb, dandelion, thistle, a new generation of flowers, wandering past the dying ranks of their parents. A few late insects dart in and out of the lengthening shadows. Columns of gnats dance in the clearings, and large Aeshna dragonflies patrol the forest glades, wings glinting in the weak sunlight.

A browning is creeping over the wood: the herbs are shrivelling and dying, their stems twisting as if to mimic the contorted oaks above. Slowly, a carpet of russet leaves is advancing, smothering the last traces of summer. Along the rides and glades, little fires appear to spread through the forest as the low afternoon sun glows through the tumbledown bracken fronds.

Autumn is a time of death, but also of renewal. Worn-out parts are being dismantled: essential nutrients stream back into roots and stems for storage through the winter. Spent leaves spiral tortuously to the ground, returning to the dust from which they sprang. Beneath this thickening shroud, bulbs and corms, swollen with food from long-dead summer leaves, soak up the last minerals of autumn with rootlets already proliferating ready for the spring burst of life.

As the moisture-laden air hangs around the tree trunks, gleaming toadstools appear around their boles, embedded in the furrowed bark and displayed on delicate pedestals rising from the leaf litter, a mysterious flora which seems to spring, fully formed, from the dead shapes of summer. From a million tiny slits under the toadstool caps, floats a life-bearing powder of spores, dusting the forest from floor to canopy.

These colourful displays are signs of a vast underground network of fine white threads, creeping across the undersides of the damp leaves and penetrating the finest interstices of fallen tree trunks. Among this world of fungal cobwebs, the process of regeneration is beginning: the corpses of this and previous years are slowly digested, and as the fungus in its turn decays, nutrients seep back into the soil to feed another generation of woodland plants.

Soon, the haze of the changing season is dispelled by autumn storms. From

---

Despite the somnolent reputation of the **dormouse** (*Muscardinus avellanarius*), when the need arises it is capable of a fair turn of speed in its arboreal habitat, moving amongst the branches with surprising agility, and even jumping at times. Unfortunately this delightful creature is much less common now.

The glorious colours of autumn leaves are no mere happy accident of nature. Having spent the summer building up its tissues, the tree salvages what it can before shedding its leaves; the pigments are broken down and the soluble products are transported back to the trunk for storage. The remainder of the degraded pigments give rise to the glorious hues of autumn. **Maples** (*Acer* sp.) are particularly brilliant at this time of year (above).

The **common frog** (*Rana temporaria*) may be found in woods where conditions are suitably moist, particularly if there are streams, ditches and pools. Almost any small patch of water will do for laying her eggs, which may number as many as two thousand.

Despite gun and trap, the **stoat** (*Mustela erminea*) has survived (left). It is largely nocturnal, although not exclusively so, hunting for any weaker animal – including rabbits and fish, which most of the weasel family especially like. The stoat has a fascinating, snake-like way of getting about, moving in a succession of undulating, low bounds. Like the fox, it may hypnotize rabbits before striking.

The forest floor is a place of dampness and darkness, inhabited not only by a host of tiny insects and other invertebrates – difficult to see – but also by a wealth of fungi. This **Coprinus fungus,** 1¼ inches (3 centimetres) high, may be found growing on rotting leaves (below). *Coprinus* species are prone to autodigestion, breaking down into a black, inky fluid.

With the onset of autumn, **spiders** are increasingly in evidence (below). They have grown during summer, and so are easier to see; also their webs are now more visible, picked out by the dew and the lower angle of the sun.

Although widespread in mainland Europe, the **little owl** (*Athene noctua*) was not introduced to Britain until the end of the last century (right). Now it is common on farmland and at the edges of woods. Little owls can often be seen sitting on wires and posts in daylight; much of their hunting takes place around dawn and dusk. They feed on insects, worms, birds and small mammals. This particular bird usually roosts in a hollow yew tree.

Seen close-to, this fungus (below) gives an impression of high glaze and fragility, hence its name: **porcelain fungus** (*Oudemansiella mucida*). It can be found from late summer through autumn high on the trunks of beech trees, in large clusters.

The **oak** (*Quercus robur*) sustains a greater variety of wildlife than any other European species of tree (right). Birds and squirrels nest in the canopy and insects devour its leaves. Fungi, algae, lichen and mosses invade the bark; birds, insects and mammals feed on the acorns, while even the roots are sought out by many small organisms. When the tree dies, its rotting carcass will continue to support a host of living creatures.

Hoverflies (*Syrphus balteatus*) are among the most colourful of flies, and there are several hundred species throughout Europe (above). The males have the ability to hover while courting the females, although the aerodynamic principles which enable them to do this are still a mystery. Hoverflies are very common throughout summer and early autumn and may be seen visiting flowers for nectar and pollen. The larvae of many are prolific aphid eaters.

The cheerful little flowers of the **herb Robert** (*Geranium robertianum*) brighten the woodland until as late as October (right). They can also be found growing on country and city walls. The flowers develop into the long, beak-shaped fruits characteristic of the storksbill family.

Another name for the **muntjac** (*Muntiacus reevesi*) is the barking deer, as it utters a sharp bark when alarmed and during the rut. Introduced from China at the turn of the century, the muntjac is a small species of the deer family, barely reaching 1½ feet (half a metre) in height. The male's antlers are short, while the upper canine teeth are so large that they project from the mouth as small tusks. The animal is more likely to be heard than seen, much preferring dense woodland to open countryside.

Brought over from North America towards the end of the last century, the **grey squirrel** (*Sciurus carolinensis*) has now colonized the British countryside (top). Although it looks appealing, its spread has coincided with the decline of the far more attractive native red squirrel (*Sciurus vulgaris*). The grey squirrel plays havoc with eggs and nestlings, and consumes oak buds, shoots, leaves, pollen, bark and acorns. In autumn it is burying fruits for the winter.

A common fungus on stumps and logs is the **tripe fungus** (*Auricularia mesenterica*), with its hairy upper surface (below). It often occurs in dense tiers.

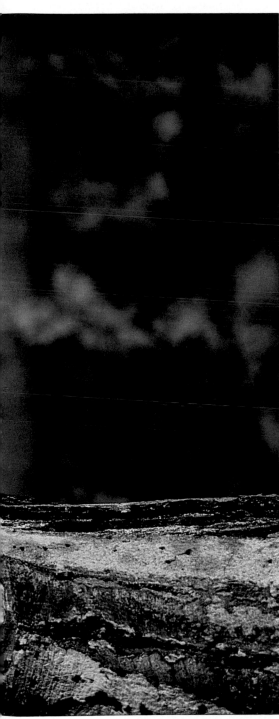

Harvestmen (*Opiliones* sp.) are related to spiders, but are characterized by very long, thin legs and a small, oval body, which has no obvious segments (left, below). Unlike spiders, harvestmen are omnivorous, feeding on the remains of dead animals, bird droppings and vegetable material, as well as hunting smaller invertebrates. They are generally nocturnal, sheltering under stones and amongst low vegetation during the day. Again unlike spiders, the female possesses a long ovipositor, with which she lays her eggs in crevices, moss or rotten wood.

Ivy (*Hedera helix*) must be one of the most familiar of plants; it not only abounds in deciduous woods, climbing up the boles of trees, but often smothers walls and rocks, and carpets the ground under hedgerows. Ivy leaves are very variable: the well-known five-lobed type grow on non-flowering shoots, and the pointed non-lobed variety on bushy flowering branches. Its flowers, which appear between October and November, seem irresistible to the local insect population, often attracting flies, bees and butterflies in swarms.

Mosses are primitive plants and are much more in evidence during the cold months when the woodland has scant vegetation; in fact winter is their most active growing time, as moisture is so abundant. Mosses lack true leaves, stems, roots and the xylem vessels which conventional plants possess for conducting food and water upwards. Instead, they absorb water over their entire surface, every cell being in contact with the growing area to obtain its water supply. In spring they produce spore capsules, and as the summer progresses they wither and dry, the cycle being repeated when cool damp conditions return in autumn.

The spores of this **star moss** (*Polytrichum* sp.) are shaken out between tiny teeth at the top of the capsule when the capsule is disturbed by rain, wind or animals (left). Mosses are frequently the first plants to colonize poor soil, helping to stabilize the ground and trap the water which helps other plants to grow.

Voles are appealing little rodents and can be distinguished from mice by their much shorter tail, blunt oval head and short, round ears. Like most voles, the **field vole** (*Microtus agrestis*) feeds on roots, bulbs, fruit, seeds, snails and insects (above). Their favourite habitats are meadows and pastures, but they will also venture into gardens, hedgerows and the edges of woods, and when present in large numbers, cause considerable damage to seedlings and young trees. As it does not hibernate, the vole does not as a rule lay up food stores.

Of all the commoner mammals, the **badger** (*Meles meles*) must be the most infrequently seen, largely due to its strictly nocturnal way of life and shy nature: over three hundred camera hours were devoted to obtaining this picture. It is omnivorous, feeding on insects, grubs, roots, bulbs, birds' eggs and weak rabbits and birds. In late autumn, it prepares its deep winter chamber, bedding it with leaves which ferment and generate a moist warmth. The passages are blocked up to keep out the cold and unwelcome visitors. Some setts are still active after hundreds of years; they are kept meticulously clean.

Lacewings, together with snakeflies, have large, flimsy, net-like wings and a weak, slow flight (left, below). This lovely, pale bluish-green insect (*Chrysopa* sp.) is difficult to confuse with anything else as it flutters from one plant to another. During autumn it often comes indoors to hibernate, when it turns brown. Both adults and larvae feed on aphids. The larvae, after sucking their prey dry, cast the empty skins on to their backs for camouflage.

As autumn progresses, a number of insects are still taking advantage of the weakening sunshine. The **seven-spot ladybird** (*Coccinella 7-punctata*) is one of the largest and commonest of numerous European species of the insect (right, below). The adults and larvae consume aphids in vast quantities.

In some ways, the **dormouse** (*Muscardinus avellanarius*) seems much like a squirrel (right). However, it has a closer affinity with mice, spending much of its life in deep slumber, remaining asleep during the daylight hours, and passing the period from September to April in an exceptionally long and profound state of hibernation. It starts preparing for this in early autumn by accumulating layers of fat beneath its warm coat; this is promoted by a rich diet of seeds, berries, nuts and juicy insects, supplemented perhaps with an occasional late nestling.

One of the more irksome insects encountered in damp forests is the **mosquito** (family Culicidae), for it breeds readily in any woodland puddle or pool (below). With their complex, needle-like mouthparts, the adults suck the nectar of flowers. Only the females pierce the skin of mammals to suck blood, and thus gain valuable nutrients for their developing eggs. In autumn, having mated, the females look for suitable places to hibernate, such as hollow trees, cellars and outbuildings. Mosquitoes beat their wings about six hundred times a second.

The **southern Aeshna** (*Aeshna cyanea*) is one of the commonest large dragonflies, on the wing from July to October (right). It often travels far from water, patrolling woodland rides right up to late dusk. With an ancestry dating back around three hundred million years and a 'flight motor' which is considered primitive by some, dragonflies nonetheless give a spectacular aerial performance – fast, powerful and manoeuvrable as they hunt for flying insects.

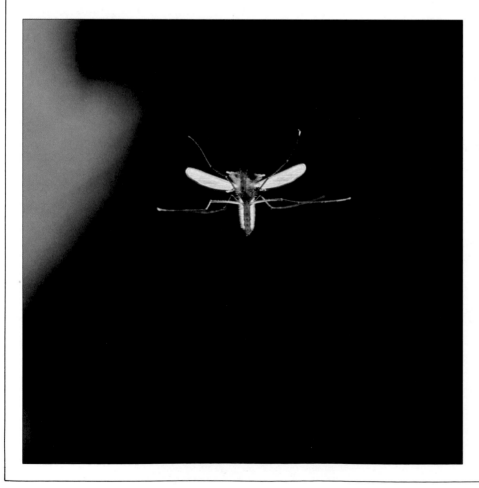

The misty days of autumn are enlivened by the fluttering of leaves twisting a path to the forest floor. In the darkness below, tangled threads of **fungal mycelium** start the slow process of decay, releasing precious nutrients into the soil (left, below).

Here, sodden **aspen** leaves (*Populus tremula*) await their turn to be transformed to soil once again (right, below).

Pure beechwoods do not support much plant and animal life, mainly because of their dense canopy and the deep layer of leaf litter. Mixed beechwoods, or better still, oakwoods with clusters of **beech** trees (*Fagus sylvatica*), are far more hospitable to wildlife, allowing light to reach the ground and encouraging many kinds of wild flowers to grow (right). Renowned for their autumn colouring, beech leaves start turning to pale yellows in early October, changing to shades of orange-brown and rufus before falling in early November.

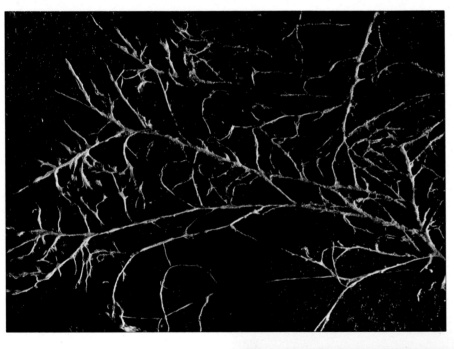

WINTER

s the year fades to a close the warm colours of autumn pale, and in the dwindling light the giant oaks reflect a steely grey, defiant against the advances of winter. Shadows deepen to purple, and the lowering skies envelop the leafless crowns as the darkest season closes in.

Now comes the rain: no longer a soft grey drizzle, but icy, penetrating darts of water, beating down the last few herbs of the woodland glades. Small woodland creatures hide in darkened hollows, trying to keep the searing wind from chilling their slender bodies.

On drier days in early winter, late insects hum as they sip the nectar of the small green flowers hidden in the shiny sheaths of ivy, and hoverflies dart through the shafts of winter sunlight. Flocks of fieldfares and redwings invade the forest in search of berries, stripping the last fruits from bushes and trees.

Although the plants are now divested of their colour, the birds already wear their courtship plumage. Blue-tits dart in the canopy, and rose-pink chaffinches bounce among the dead leaves. White-cheeked great-tits tug at grubs on oak twigs, and gold-barred greenfinches flit through the trees in search of seeds.

On the old furrowed trunks lichens are exposed, hanging like locks of greying hair, spreading over the bark in orange-fringed circles and wrapping twigs in crusty coats; delicate patterns etch across the ancient roots.

Beneath the carpet of wet leaves, stores lie hidden: seeds, nuts, acorns, and countless tiny creatures of the soil. As the trees above are plundered, the life of the wood centres on the ground below. A pigmy shrew riffles through the leaves; voles and field mice scuffle in the undergrowth; and squirrels leave the treetops to sniff for buried acorns.

For some, the search is over. Snug in a nest of dry grass in a hollow in the old oak the dormouse lies, cold and unmoving, its breathing shallow, its heartbeat slow, gradually drawing on its reserves of fat. Beneath the twisted roots the hedgehog sleeps in its bed of leaves, and deeper still the adder. On frosty nights wrens huddle close together in old nests, sometimes in astonishing numbers.

The first frosts of winter bring down even the green leaves from the trees, creating a kaleidoscope of green, yellow and brown on the forest floor.

Before falling in October, birch leaves turn bright yellow. Here, a **birch** leaf (*Betula pendula*) has blown on to a mat of moss which even in the dead of winter vibrates green with life (below). Most mosses grow actively during the cool, moist winter months, and are particularly luxuriant at the end of the season.

In autumn the **dormouse** (*Muscardinus avellanarius*) constructs a nest under moss, dead leaves or tree roots, where it hibernates for six months of the year (right). The sleep is profound, the heart rate and respiration fall, and the body temperature becomes so low that the animal feels cold to the touch and is difficult to rouse. The low rate of metabolism is maintained by very slowly using up fat accumulated during the autumn.

Although flowerless, **lichens** are beautifully coloured in a variety of subtle shades of green (left). Many are epiphytic, absorbing nutrients from the surrounding air rather than from the soil as in conventional plants. Like mosses, they absorb water all over their surface, so thrive particularly well in areas saturated in water vapour, often festooning the trunks and branches of trees. Sadly, because of pollution lichens are much less common than they used to be.

On moisture-laden winter nights, the hoar frost traces crystal lines around the relics of the summer (above).

Bold patterns of branch and twig
are highlighted by a winter snowfall
(below).

In early winter the **blackbird** (*Turdus
merula*) and its relatives, the redwing
(*Turdus muscicus*) and fieldfare
(*Turdus pilaris*), feast on the remainder
of the autumn berries and other soft
fruits (right). Even in the depths of
winter it can usually find a good
supply of worms, which live nearer
the surface when the soil is moist.

The confident song of the **robin** (*Erithacus rubecula*) is one of the first in winter to penetrate the cold air (left). Despite its seeming exuberance, the song is really a defiant proclamation of the singer's territorial rights.

Of all the smaller mammals, the **hedgehog** or hedgepig (*Erinaceus europaeus*) must be one of the most popular (below). Being so well protected, it has few natural enemies, although foxes and badgers sometimes succeed in unrolling the ball of spines. Hedgehogs are extremely noisy creatures, with a range of some twenty-five different grunts, squeaks and screams. The males can frequently be heard quarrelling at night.

As their eyesight is poor, hedgehogs use their acute sense of smell for locating food; this consists of slugs, snails, insects, birds' eggs and occasionally larger creatures such as frogs, mice and even snakes. Surprisingly, they are excellent swimmers and will climb walls and trees with relative ease. Most of the cold months are spent in hibernation, snuggled up in a bed of dry leaves and moss in the hole of a bank or under the roots of a tree; however, during mild spells, particularly in early winter, the hedgehog wakes up for brief phases of activity.

Beneath the dense leaf litter lie the **bluebell** bulbs (*Hyacinthoides non-scriptus*), with all their stores of energy from the previous summer's sun, ready to be converted into the first leaves of the new season (left, above and below). Already, shoots are pushing up through the litter, having passed the winter well insulated by the leaves and the occasional coat of snow.

As spring approaches, the **feather moss** (*Thuidium tamariscinum*) is growing bright new fronds to replace the dull brown relics of winter (right).

Now that winter has passed by, a misty-green, luminescent glow appears amidst the grey woodland landscape as the first **birch** leaves (*Betula pendula*) unfurl once more (overleaf).

Notes on the Photography

Lighting

When working in the field I like to make use of daylight whenever possible: the beauty and subtlety of natural light is difficult if not impossible to match with flash. Thus over two-thirds of the photographs that appear in this book were exposed with natural light. This includes all the shots of plants and fungi (except spore dispersal), of reptiles and of some of the less active birds, mammals and insects.

There are of course occasions when flash is indispensible, for instance when working in the dark or at night and when it is necessary to arrest rapid movement. To freeze the high-speed activities of flying birds, insects and leaping mice, I employed special flash equipment with a speed of about 1/20,000th of a second, together with the associated light beams and electronic gadgetry for detecting the creatures at the right spot, and firing the camera at the right time. Many of these subjects were captured on film in their natural habitat, although for practical reasons the flying insects had to be photographed inside.

Film

Film was chosen to provide maximum detail and minimum grain. On a 35mm format, Kodachrome 25 (and occasionally Kodachrome 64) was used, and Fujichrome 50 on the 2¼ inch square format.

Equipment

For most field work I use a Nikon F3 body with 24mm, 55mm, 105mm Nikkor lenses and occasionally a 300mm IF ED lens. Many high-speed photographs of insects were taken with an old Leicaflex SL, equipped with a twenty-five-year-old Leitz 135mm Hektor. Some of the general close-ups, including the insects at rest, were shot with a Leitz 100mm Macro Elmar attached to the Nikon F3 via a Novoflex bellows! A Hasselblad ELM with a 150mm Sonnar or 50mm Distagon was used for some of the landscapes, birds and mammals.

Of all the photographs reproduced here, only two were exposed without a tripod: I leave you to guess which ones.

Equipment alone does not produce good results, although reliable cameras with good optics can help. The less paraphernalia I carry around the countryside, the more pictures I seem to take. Apart from the obvious weight problem, less equipment means fewer time-consuming decisions. The trick is to know what to leave at home. By far the most valuable prerequisites for nature photography are an understanding and feeling for the subject; observing plants and animals within their surroundings, watching the effects of light and shade, being patient and sustaining the right frame of mind, are the real secrets.

Index

A

Accipiter nisus, 93
Acer, 114
adder, 79
Adela viridella, 59
Aesculus hippocastanum, 42
Aeshna cyanea, 137
Agrion virgo, 91
Allium ursinum, 61
Amanita muscaria, 109
Anemone nemorosa, 24
Anthocharis cardamines, 57
Apis mellifera, 97
Apodemus sylvaticus, 49, 60
aspen, 138
Argynnis paphia, 81
Athene noctua, 119
Auricularia mesenterica, 125

B

badger, 133
banded snail, 113
barking deer, 124
beech, 50, 51, 139
Betula pendula, 31, 146, 156–7
birch, 31, 146, 156–7
bird's-eye speedwell, 69
blackbird, 151
blackcap, 78
bluebell, 29, 45, 154
blue-tit, 58
Boloria selene, 92
Bombus, 73
bracken, 41
bramble, 26, 27
brimstone butterfly, 28
brown owl, 34–5
brown rat, 33
buckthorn, 28
bullfinch, 40
bumblebee, 73

C

cabbage white butterfly, 28
Caltha palustris, 36
Cepaea nemoralis, 113
Capreolus capreolus, 43
Cardamine pratensis, 36

cardinal beetle, 74
Carex pendula, 100
chestnut, 42
chicken-of-the-woods, 100
Chrysopa, 134
Chrysosplenium oppositifolium, 37
Cladius viminalis, 74
click beetle, 80
Coccinella 7-punctata, 134
Columba palumbus, 52
common agrion, 91
common helleborine, 88
Coprinus fungus, 117
Corvus monedula, 44
Corylus avellana, 24
crane-fly, 23
cuckoo flower, 36, 57
Cuculio nucum, 75
Culicidae, 136

D

Dactylorhiza fuchsii, 89
Daddy-long-legs, 23
Deilephila elpenor, 76
Dendrocopos major, 82
Digitalis purpurea, 72
Dioctrea linearis, 75
dog rose, 83
dormouse, 111, 135, 147
drooping sedge, 100

E

Elateridae, 80
elephant hawk moth, 76
Epilobium angustifolium, 96, 97
Epipactis helleborine, 88
Erinaceus europaeus, 153
Erithacus rubecula, 152
Euphorbia amygdaloides, 100

F

Fagus sylvatica, 50, 51, 139
feather moss, 54, 154
ferns, 31
field vole, 132
fireweed, 96, 97

fly agaric, 109
fox, 38
foxglove, 72
frog, 115
frost, 149
fungal mycelium, 138

G

Garrulus glandarius, 67, 85
Geranium robertianum, 123
germander speedwell, 69
giant lacewing, 86, 87
golden saxifrage, 37
Gonepteryx rhamni, 28
grass snake, 37
greater stitchwort, 56
great spotted woodpecker, 82
green longhorn, 59
green oak-roller, 48
green-veined white butterfly, 46
grey squirrel, 125
guelder rose, 90

H

harvestman, 126
hazel, 24
Hedera helix, 127
hedgehog, 153
herb Robert, 123
honeybee, 97
honeysuckle, 42, 94
hoverfly, 122
Hyacinthoides non-scriptus, 29, 45, 154

I

ivy, 127

J

jackdaw, 44
jay, 67, 85

K

kingcup, 36

L

lacewing, 134
Lactiporus sulphureus, 101
lady's smock, 36
Leptophyes punctatissima, 80
lichen, 148
Limenitis camilla, 95
Linyphiidae, 47
little owl, 119
long-eared bat, 105, 128
long-tailed field mouse, 49, 60
Lonicera periclymenum, 42, 94
Lychnis flos-cuculi, 88
Lycoperdon, 129
Lysimachia nemorum, 104

M

maple, 114
marsh marigold, 36
marsh tit, 23
Meles meles, 38, 133
Microtus agrestis, 132
money spider, 47
mosquito, 136
Muntiacus reevesi, 124
muntjac, 124
Muscardinus avellanarius, 111, 135, 147
Mustela erminea, 116
mycelium, 138

N

Natrix natrix, 37
Noctua pronuba, 128
nut weevil, 75
nuthatch, 102

O

oak, 19, 99, 105, 121, 149
Opiliones, 126
orange-tip butterfly, 57
Osmylus fulvicephalus, 86, 87
Oudemansiella mucida, 120
Oxalis acetosella, 55

P

painted lady, 65
Pararge aegeria, 70
Parus caeruleus, 58

Parus montanus, 23
pendulous sedge, 100
Phasianus colchicus, 38
pheasant, 38
Phylloscopus trochilus, 71
Pieris brassicae, 28
Pieris napi, 46
pipistrelle bat, 145
Pipistrellus pipistrellus, 145
Plecotus auritus, 105, 128
Polypodiaceae, 31
Polytrichum, 131
poplar sawfly, 74
Populus tremula, 138
porcelain fungus, 120
privet hawk moth, 77
Pteridium aquilium, 41
puffball, 129
Pyrochroa serraticornis, 74
Pyrrhula pyrrhula, 40

Q

Quercus robur, 19, 99, 105, 121, 149

R

ragged robin, 88
ramsons, 61
Rana temporaria, 115
Rattus norvegicus, 33
red campion, 53
Rhamnus catharticus, 28
robberfly, 75
robin, 152
roe deer, 43
Rosa canina, 83
rosebay willowherb, 96, 97
Rubus fruticosus, 26, 27

S

Sciurus carolinensis, 125
seedlings, 30
seven-spot ladybird, 134
Silene dioica, 53
silver birch, 31, 146
silver-washed fritillary, 81
Sitta europaea, 102
small pearl-bordered fritillary, 92
southern Aeshna, 137
sparrowhawk, 93
speckled bush cricket, 80
speckled wood butterfly, 70

Sphagnum moss, 32
Sphinx ligustri, 76
spider, 118
spotted longhorn, 103
spotted orchis, 89
star moss, 131
Stellaria holostea, 56
stitchwort, 56
stoat, 116
Strangalia maculata, 103
Streptopelia turtur, 98
Strix aluco, 34–5
sulphur polypore, 101
Sylvia atricapella, 78
Syrphus balteatus, 122

T

tawny owl, 34–5
Thuidium tamariscinum, 54, 154
Tipulidae, 23
Tortrix viridana, 48
tripe fungus, 125
Troglodytes troglodytes, 26
Turdus merula, 151
turtle dove, 98

V

Vanessa cardui, 65
Veronica chamaedrys, 69
Viburnum opulus, 90
Vipera berus, 79
Vulpes vulpes, 38

W

white admiral, 95
willow tit, 21
willow warbler, 71
winter snowfall, 150
wood anemone, 24
wood garlic, 61
wood mouse, 49, 60
wood pigeon, 52
wood sorrel, 55
wood spurge, 100
wren, 26

Y

yellow pimpernel, 104
yellow underwing moth, 128

Index

A

Accipiter nisus, 93
Acer, 114
adder, 79
Adela viridella, 59
Aesculus hippocastanum, 42
Aeshna cyanea, 137
Agrion virgo, 91
Allium ursinum, 61
Amanita muscaria, 109
Anemone nemorosa, 24
Anthocharis cardamines, 57
Apis mellifera, 97
Apodemus sylvaticus, 49, 60
aspen, 138
Argynnis paphia, 81
Athene noctua, 119
Auricularia mesenterica, 125

B

badger, 133
banded snail, 113
barking deer, 124
beech, 50, 51, 139
Betula pendula, 31, 146, 156–7
birch, 31, 146, 156–7
bird's-eye speedwell, 69
blackbird, 151
blackcap, 78
bluebell, 29, 45, 154
blue-tit, 58
Boloria selene, 92
Bombus, 73
bracken, 41
bramble, 26, 27
brimstone butterfly, 28
brown owl, 34–5
brown rat, 33
buckthorn, 28
bullfinch, 40
bumblebee, 73

C

cabbage white butterfly, 28
Caltha palustris, 36
Cepaea nemoralis, 113
Capreolus capreolus, 43
Cardamine pratensis, 36

cardinal beetle, 74
Carex pendula, 100
chestnut, 42
chicken-of-the-woods, 100
Chrysopa, 134
Chrysosplenium oppositifolium, 37
Cladius viminalis, 74
click beetle, 80
Coccinella 7-punctata, 134
Columba palumbus, 52
common agrion, 91
common helleborine, 88
Coprinus fungus, 117
Corvus monedula, 44
Corylus avellana, 24
crane-fly, 23
cuckoo flower, 36, 57
Cuculio nucum, 75
Culicidae, 136

D

Dactylorhiza fuchsii, 89
Daddy-long-legs, 23
Deilephila elpenor, 76
Dendrocopos major, 82
Digitalis purpurea, 72
Dioctrea linearis, 75
dog rose, 83
dormouse, 111, 135, 147
drooping sedge, 100

E

Elateridae, 80
elephant hawk moth, 76
Epilobium angustifolium, 96, 97
Epipactis helleborine, 88
Erinaceus europaeus, 153
Erithacus rubecula, 152
Euphorbia amygdaloides, 100

F

Fagus sylvatica, 50, 51, 139
feather moss, 54, 154
ferns, 31
field vole, 132
fireweed, 96, 97

fly agaric, 109
fox, 38
foxglove, 72
frog, 115
frost, 149
fungal mycelium, 138

G

Garrulus glandarius, 67, 85
Geranium robertianum, 123
germander speedwell, 69
giant lacewing, 86, 87
golden saxifrage, 37
Gonepteryx rhamni, 28
grass snake, 37
greater stitchwort, 56
great spotted woodpecker, 82
green longhorn, 59
green oak-roller, 48
green-veined white butterfly, 46
grey squirrel, 125
guelder rose, 90

H

harvestman, 126
hazel, 24
Hedera helix, 127
hedgehog, 153
herb Robert, 123
honeybee, 97
honeysuckle, 42, 94
hoverfly, 122
Hyacinthoides non-scriptus, 29, 45, 154

I

ivy, 127

J

jackdaw, 44
jay, 67, 85

K

kingcup, 36

L

lacewing, 134
Lactiporus sulphureus, 101
lady's smock, 36
Leptophyes punctatissima, 80
lichen, 148
Limenitis camilla, 95
Linyphiidae, 47
little owl, 119
long-eared bat, 105, 128
long-tailed field mouse, 49, 60
Lonicera periclymenum, 42, 94
Lychnis flos-cuculi, 88
Lycoperdon, 129
Lysimachia nemorum, 104

M

maple, 114
marsh marigold, 36
marsh tit, 23
Meles meles, 38, 133
Microtus agrestis, 132
money spider, 47
mosquito, 136
Muntiacus reevesi, 124
muntjac, 124
Muscardinus avellanarius, 111, 135, 147
Mustela erminea, 116
mycelium, 138

N

Natrix natrix, 37
Noctua pronuba, 128
nut weevil, 75
nuthatch, 102

O

oak, 19, 99, 105, 121, 149
Opiliones, 126
orange-tip butterfly, 57
Osmylus fulvicephalus, 86, 87
Oudemansiella mucida, 120
Oxalis acetosella, 55

P

painted lady, 65
Pararge aegeria, 70
Parus caeruleus, 58

Parus montanus, 23
pendulous sedge, 100
Phasianus colchicus, 38
pheasant, 38
Phylloscopus trochilus, 71
Pieris brassicae, 28
Pieris napi, 46
pipistrelle bat, 145
Pipistrellus pipistrellus, 145
Plecotus auritus, 105, 128
Polypodiaceae, 31
Polytrichum, 131
poplar sawfly, 74
Populus tremula, 138
porcelain fungus, 120
privet hawk moth, 77
Pteridium aquilium, 41
puffball, 129
Pyrochroa serraticornis, 74
Pyrrhula pyrrhula, 40

Q

Quercus robur, 19, 99, 105, 121, 149

R

ragged robin, 88
ramsons, 61
Rana temporaria, 115
Rattus norvegicus, 33
red campion, 53
Rhamnus catharticus, 28
robberfly, 75
robin, 152
roe deer, 43
Rosa canina, 83
rosebay willowherb, 96, 97
Rubus fruticosus, 26, 27

S

Sciurus carolinensis, 125
seedlings, 30
seven-spot ladybird, 134
Silene dioica, 53
silver birch, 31, 146
silver-washed fritillary, 81
Sitta europaea, 102
small pearl-bordered fritillary, 92
southern Aeshna, 137
sparrowhawk, 93
speckled bush cricket, 80
speckled wood butterfly, 70

Sphagnum moss, 32
Sphinx ligustri, 76
spider, 118
spotted longhorn, 103
spotted orchis, 89
star moss, 131
Stellaria holostea, 56
stitchwort, 56
stoat, 116
Strangalia maculata, 103
Streptopelia turtur, 98
Strix aluco, 34–5
sulphur polypore, 101
Sylvia atricapella, 78
Syrphus balteatus, 122

T

tawny owl, 34–5
Thuidium tamariscinum, 54, 154
Tipulidae, 23
Tortrix viridana, 48
tripe fungus, 125
Troglodytes troglodytes, 26
Turdus merula, 151
turtle dove, 98

V

Vanessa cardui, 65
Veronica chamaedrys, 69
Viburnum opulus, 90
Vipera berus, 79
Vulpes vulpes, 38

W

white admiral, 95
willow tit, 21
willow warbler, 71
winter snowfall, 150
wood anemone, 24
wood garlic, 61
wood mouse, 49, 60
wood pigeon, 52
wood sorrel, 55
wood spurge, 100
wren, 26

Y

yellow pimpernel, 104
yellow underwing moth, 128